Generously donated by
The Goodwyn Institute

MEMPHIS LIBRARY FOUNDATION

GREEN MATTERS™

Making Good Choices About
RECYCLING AND REUSE

STEPHANIE WATSON

rosen publishing's
rosen
central®

To Jake and his friends at the Galloway School,
for their commitment to recycling

Published in 2010 by The Rosen Publishing Group, Inc.
29 East 21st Street, New York, NY 10010

First Edition

Library of Congress Cataloging-in-Publication Data

Watson, Stephanie, 1969–
Making good choices about recycling and reuse / Stephanie Watson.—1st ed.
     p. cm.—(Green matters)
Includes bibliographical references and index.
ISBN-13: 978-1-4358-5312-6 (library binding)
ISBN-13: 978-1-4358-5606-6 (pbk)
ISBN-13: 978-1-4358-5607-3 (6 pack)
1. Recycling (Waste, etc.)—Juvenile literature. 2. Waste products—Juvenile literature.
I. Title.
TD794.5.W3828 2010
363.72'82—dc22

                                                                    2008046094

Manufactured in Malaysia

CONTENTS

INTRODUCTION

How much trash do you throw away? If you're like the average American, you toss more than 4 pounds (just under 2 kilograms) of garbage every day of the year. That's just you—one person. Add in the more than 300 million people living in this country, and it adds up to 250 million tons (227 million metric tons) of garbage produced each year, according to the U.S. Environmental Protection Agency's (EPA) *2006 Facts and Figures*. These are just a few of the items that Americans throw away each year, according to the EPA: 25 billion Styrofoam cups, 2 billion used batteries, 15 million tons (about 14 million metric tons) of food waste, 1 billion fruit juice boxes, 16 billion disposable diapers, and 700,000 old television sets.

This list merely scratches the surface of all the trash in the United States. Add to it food scraps, junk mail, cell phones, bicycles, books, and literally thousands of other consumer goods, and you can see why the country is dealing with a whole lot of trash.

WHAT TO DO WITH SO MUCH TRASH

Imagine if all the trash created every year—all 250 million tons of it—were just piled up on the street. It would stretch high into the sky, creating skyscrapers of garbage. The smell from rotten food and waste products would be overpowering.

Fortunately, Americans have places to put their trash. When you or your parents roll the garbage bin out to the curb once or twice every week, a garbage truck picks it up and takes it to a transfer station.

From there, most garbage goes to a landfill—a large, deep area in the ground where trash is deposited, layer upon layer.

You don't have to worry about trash piling up on your street, which is good news. The bad news is that sticking tons and tons of waste in a landfill means that all those pieces of garbage—from tennis shoes to tires—can never be used again.

What if, instead of throwing away all those old items, you turned them into something else? What if you used your old food scraps to make fertilizer for your garden or flower bed? What if you transformed an old CD case into a picture frame for your desk? Turning something old and unneeded into something new and needed is the idea behind recycling and reuse.

THE THREE R'S

Recycling is just one part of the "three R's"—reduce, reuse, and recycle. All of the "R's" represent ways to conserve natural resources.

"Reduce" means using less by choosing products that are long lasting. Instead of eating your dinner on disposable plates and then throwing them away, you use regular dishes and wash them when you're finished. Rather than packing your lunch in a brown paper bag, you bring the same lunchbox or canvas bag to school every day. You can also bring a reusable cup instead of using and throwing away disposable cups.

This consumer living in Oakland, California, kept all his trash for one year to prove that he could cut down on the amount of disposable goods that he used; he eventually reached his goal of not needing to throw anything away.

"Reuse" means finding another use for items instead of throwing them away. You might donate your old sunglasses to a charity or have your broken computer fixed so that you don't have to trade it in for a new one.

"Recycle" means turning materials you're no longer using into something you can use. You could recycle old newspapers to be turned into new newsprint or give your empty soda cans to a recycling center so that they can be melted down and turned into new cans. Unlike reusing, recycling breaks down old products into raw materials, which are then made into new items.

The idea behind recycling is very simple, and there are many ways to recycle. Most communities have recycling programs to make the process easy. All you have to do is sort your used items (paper, glass, aluminum) and a truck will pick them up and take them to a recycling center. Sometimes, you can even earn money by turning in used bottles and cans.

Recycling isn't a new idea, but it has been catching on in recent years. In 1960, Americans recycled just 6 percent of the products they used. Today, that number is up to more than 32 percent, according to the EPA. The increase in recycling is great news, but people are still throwing away almost 70 percent of their garbage. Recycling in America has a long way to go.

If people can increase the amount of waste they recycle, they can conserve more of the world's natural resources. A good example is trees. For every ton of paper recycled, seventeen trees are saved. What's more, many of the toxic chemicals that are normally used in the papermaking process are not released into the environment.

In this book, you'll learn how recycling works and see the many ways it helps conserve the earth's natural resources. You'll follow some of the items you use every day as they are transformed into new products. And you'll see why some people think recycling is a great idea, while others think it's not.

This book should leave you with some ideas on how to start reusing and recycling at home and at school. Try them out and show your parents and teachers how they can help. The next time you head to the trash, stop. Look at the item in your hand and think about whether you can put it in your recycling bin or find another use for it. It can be your first recycling project!

CHAPTER ①

Recycling Basics

To avoid towering skyscrapers of piled-up trash everywhere, people have come up with a variety of ways to dispose of the vast amounts of garbage they produce. The three main methods for getting rid of trash are landfills, incineration, and recycling.

Most of the trash produced by households goes into large areas called landfills (also called dumps). Garbage that is collected from one town or city is transported by truck or train to a great big hole in the ground. Wastes are released into sections of the landfill called cells. About 2 feet (61 centimeters) of wet clay is put down and fills itself in as it settles over time. Thick plastic is laid over the clay. About 8 feet (2.4 meters) of waste is placed in each cell and is then covered with about 6 inches (15.2 cm) of soil called daily cover. Another layer of garbage is added, followed by another layer of dirt, and so on. The capping, or top cover, of the cell is sloped so that precipitation

The Fresh Kills Landfill on Staten Island, New York, used to be the largest landfill in the world. It once processed about 15,000 tons (13,600 metric tons) of trash each day.

will not collect on the top. Finally, grass is planted over the top cover. Filling a landfill is like making a layer cake—although a very dirty, smelly one.

It's obvious that garbage isn't clean. Many of the items thrown in the trash contain toxic chemicals, which could drain into the water supply or get into the air. (Toxic waste must be disposed of at a hazardous waste site. It is illegal to dump toxic waste at a landfill for household waste.) When these chemicals combine, they produce a liquid substance called leachate. Landfills for household waste have barriers that help slow down the release of the leachate. The bottom of the landfill is lined with compacted clay, which is then covered in a thick plastic liner. The plastic liner is not supposed to break down and thus is made to last a long time. (Currently, some researchers believe that liners delay the process of the leachate flowing out and contaminating groundwater, but they do not prevent it from doing so. Usually, the liner is checked for any leakage.) Landfills also have special drainage systems—pipes and valves that collect liquid runoff and send it to a treatment plant. At the treatment plant, toxic chemicals and heavy metals are removed. The remaining liquids are cleaned and purified, and they are then returned to the water supply.

Chemicals in landfills aren't just in liquid form; sometimes, they can be released into the air as gas. As garbage is exposed to the sun, rain, and air, it decomposes (breaks down) and releases methane gas, which can be dangerous to the environment and can easily catch fire. Landfills have special systems in place to monitor and control how much methane gas is produced. Many landfills capture methane gas and convert it into energy.

The second method of removing wastes is called incineration. Incineration can help reduce the amount of trash put in landfills. Waste is burned in a machine called an incinerator. The first incinerator was built in 1885 on Governor's Island in New York Harbor. By the early twentieth century, cities had begun using their trash incinerators to produce energy.

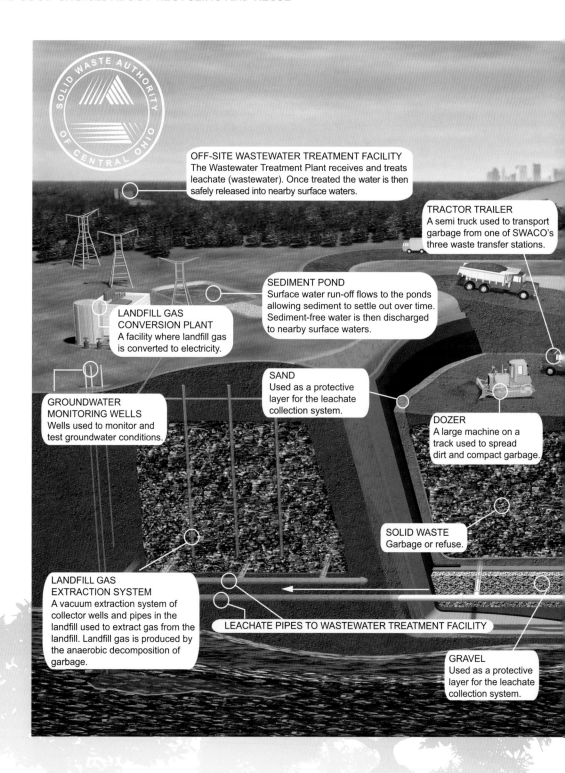

OFF-SITE WASTEWATER TREATMENT FACILITY
The Wastewater Treatment Plant receives and treats leachate (wastewater). Once treated the water is then safely released into nearby surface waters.

TRACTOR TRAILER
A semi truck used to transport garbage from one of SWACO's three waste transfer stations.

SEDIMENT POND
Surface water run-off flows to the ponds allowing sediment to settle out over time. Sediment-free water is then discharged to nearby surface waters.

LANDFILL GAS CONVERSION PLANT
A facility where landfill gas is converted to electricity.

SAND
Used as a protective layer for the leachate collection system.

GROUNDWATER MONITORING WELLS
Wells used to monitor and test groundwater conditions.

DOZER
A large machine on a track used to spread dirt and compact garbage.

SOLID WASTE
Garbage or refuse.

LANDFILL GAS EXTRACTION SYSTEM
A vacuum extraction system of collector wells and pipes in the landfill used to extract gas from the landfill. Landfill gas is produced by the anaerobic decomposition of garbage.

LEACHATE PIPES TO WASTEWATER TREATMENT FACILITY

GRAVEL
Used as a protective layer for the leachate collection system.

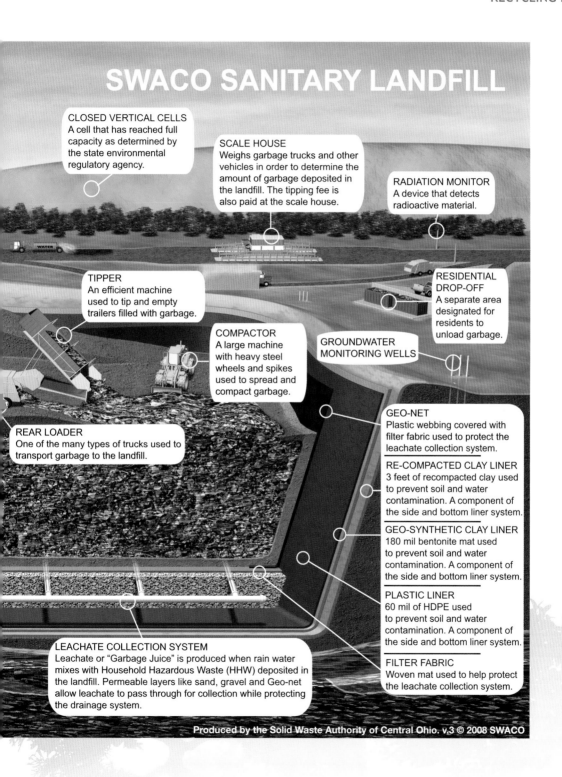

SWACO SANITARY LANDFILL

CLOSED VERTICAL CELLS
A cell that has reached full capacity as determined by the state environmental regulatory agency.

SCALE HOUSE
Weighs garbage trucks and other vehicles in order to determine the amount of garbage deposited in the landfill. The tipping fee is also paid at the scale house.

RADIATION MONITOR
A device that detects radioactive material.

TIPPER
An efficient machine used to tip and empty trailers filled with garbage.

RESIDENTIAL DROP-OFF
A separate area designated for residents to unload garbage.

COMPACTOR
A large machine with heavy steel wheels and spikes used to spread and compact garbage.

GROUNDWATER MONITORING WELLS

REAR LOADER
One of the many types of trucks used to transport garbage to the landfill.

GEO-NET
Plastic webbing covered with filter fabric used to protect the leachate collection system.

RE-COMPACTED CLAY LINER
3 feet of recompacted clay used to prevent soil and water contamination. A component of the side and bottom liner system.

GEO-SYNTHETIC CLAY LINER
180 mil bentonite mat used to prevent soil and water contamination. A component of the side and bottom liner system.

PLASTIC LINER
60 mil of HDPE used to prevent soil and water contamination. A component of the side and bottom liner system.

FILTER FABRIC
Woven mat used to help protect the leachate collection system.

LEACHATE COLLECTION SYSTEM
Leachate or "Garbage Juice" is produced when rain water mixes with Household Hazardous Waste (HHW) deposited in the landfill. Permeable layers like sand, gravel and Geo-net allow leachate to pass through for collection while protecting the drainage system.

Produced by the Solid Waste Authority of Central Ohio. v.3 © 2008 SWACO

Some cities encourage or require their residents to recycle. In Seattle, Washington, for example, residents' trash will not get picked up if it contains too many recyclable goods.

Even though incineration is an effective way to destroy trash and create energy, it is not used as often as it once was. The reason, in part, is because of concerns that burning trash can release harmful chemicals into the air. Today, only about 20 percent of solid waste is incinerated, according to the EPA.

Recycling—making garbage into something new so that it can be used again—is the third method for getting rid of trash. Just about everything, from soda cans to newspapers, printer cartridges to iPods,

BEST AND WORST CITIES FOR RECYCLING

Though recycling has become more popular in recent years, not every city is catching on. While some cities recycle most of their trash, others send the majority to landfills. A 2008 study by the trade magazine *Waste News* found that San Francisco and Los Angeles, California, were among the best cities for recycling. Which city is the worst recycler? Houston, Texas. Houston recycles just 2.6 percent of its total waste, as reported in July 2008 in the *New York Times*.

can be recycled. Sometimes, an item made from recycled products becomes more valuable than the original product. New York City artist Justin Gignac sells cubes made of garbage that he finds at Yankee Stadium. He charges $100 for each garbage cube. Designers are creating clothing and jewelry out of fibers taken from recycled textiles. These pieces often cost more than the original garments from which they were made.

RECYCLING PROGRAMS

Should people be required to recycle, or should they be able to choose whether or not they do? The answer to that question depends on where in the nation you live.

Some communities have mandatory recycling—laws or ordinances that require people to recycle. Which products are recycled depends on the local laws, but people who refuse to recycle must pay a fine.

The City View Center in Ohio is a shopping mall that was built on two former landfills.

Other areas of the country have voluntary recycling programs in which people are encouraged, but not required, to recycle. One of the ways communities encourage recycling is with pay-as-you-throw programs. These programs charge people for trash pickup based on the amount of garbage they throw away. For example, a 12-gallon (45.4 liter) trashcan may cost $10 per month for pickup, whereas a 32-gallon (121 l) trashcan costs $30 per month. People who are part of this program want to recycle because it saves them money on their trash.

Which is more effective—voluntary or mandatory recycling? According to the U.S. Department of Energy, mandatory programs are more successful. About half of the people in mandatory programs recycle, compared to only one-third of the people in voluntary programs.

REASONS TO RECYCLE AND REUSE

The biggest benefit of recycling and reusing is that they reduce the amount of trash that goes into landfills. In 2006, recycling kept 82 million tons (about 74 million metric tons) of waste out of landfills, according to the EPA. Yet, recycling didn't eliminate all the garbage that needed to be thrown away.

The United States is a big country, but it still has to house more than 301 million people—and their trash. As cities expand, space for landfills becomes harder to find. Space that is available isn't always usable because people don't want landfills close to their homes and schools.

The reason why people don't want to live near landfills is that they are often full of chemicals, such as pesticides, paint thinners, and battery acid. Although safeguards are in place to help keep those chemicals from getting into the air and water supply, some health experts say the safeguards aren't perfect. A few studies have found that people who live close to landfills have a higher risk of getting cancer than people living farther away.

Landfills also aren't free. Garbage companies must pay a tipping fee to dump their trash. The tipping fee is based on the weight of the garbage. The trash company must pass the cost along to its customers.

In addition to saving landfill space, here are a few of the other benefits of recycling:

1. Recycling saves natural resources. It conserves trees, which are essential to the environment because they pull

carbon dioxide gas from the air. Carbon dioxide is a major contributor to global warming.

2. Recycling can reduce air and water pollution. It helps keep chemicals and toxic trash out of landfills. Recycling also reduces the amount of pollutants released during the manufacture of new products. Recycling 1 ton (0.9 metric tons) of steel prevents 200 pounds (90.7 kg) of air pollutants, 100 pounds (45.4 kg) of water pollutants, and almost 3 tons (2.7 metric tons) of mining waste from being released, according to Harvey Blatt in *America's Environmental Report Card*.

3. Recycling saves energy. It takes a lot more energy to find raw materials, make a new product, and then dispose of it than it does to recycle an old product. Recycling one aluminum can, for example, uses 95 percent less energy than making new aluminum out of ore, says the EPA.

4. Recycling creates jobs. According to the National Recycling Coalition, recycling is a $236 billion-a-year industry. More than one million people have jobs collecting, transporting, and processing recycled items. Businesses are able to create and sell many new products that are made using recycled materials.

Overall, recycling is a great idea. Yet, believe it or not, there are also many convincing reasons that people give for not recycling.

REASONS NOT TO RECYCLE AND REUSE

Some critics give the following reasons for why they believe recycling isn't always a good thing to do:

1. Landfill space is abundant. U.S. landfills are nowhere near holding their capacity—there is still room to hold all the

A worker at a recycling center in Maryland picks out nonrecyclable trash. People who support recycling believe that it creates jobs. Critics of recycling believe that it uses too much fuel and energy in collecting, transporting, and processing materials to be cost-effective.

trash this country makes, and more. A. Clark Wiseman, an economist at Gonzaga University in Spokane, Washington, once calculated that all the trash that the United States creates by the year 3000 could fit in a square piece of land 100 yards (91.4 m) deep and 35 miles (56.3 km) wide on each side.

2. Landfills are safe. Modern landfills have so many precautions in place, from plastic liners to drainage systems, that some experts say there is no risk of any chemicals escaping.

3. Recycling uses energy and creates pollution. Extra trucks are needed to pick up recycling and drop it off at processing facilities. It takes energy to produce new goods from recycled ones. Both processes of picking up recyclables and turning them into new products release pollution into the air.

4. Recycling is expensive. It often costs city recycling programs more to pick up and sort bottles, cans, and other products than the cities receive when they sell those items to manufacturers.

5. People recycle more than they can use. Recycling has become so popular that some products are in abundant supply. For example, the United States sells much of its recycled paper to China because there aren't enough uses for all that paper in the United States.

6. Recycling does not work forever. Many items cannot be recycled over and over again. Paper, for example, will wear out after being recycled a few times. Eventually, manufacturers will have to cut down more trees and find other raw materials. Recycling does not save resources forever—it just delays their use.

One of the most well-known arguments against recycling was written by journalist John Tierney in a 1996 article in the *New York Times* called "Recycling Is Garbage." According to the article, although recycling has good intentions and makes sense for some materials, disposing of trash in landfills is still easier and less expensive overall. Tierney wrote, "Recycling may be the most wasteful activity in modern America: a waste of time and money, a waste of human and natural resources."

The arguments both for and against recycling are likely to continue in the years to come. In the future, it will be up to you and your peers to decide what to do with your garbage.

CHAPTER 2

What Is Recycled?

Recycling begins at the curb outside your house. You put your recyclables—plastic bottles, aluminum cans, and paper—into one bin or separate them into individual bins. About once each week, a truck comes by to pick up those items. Some recycling trucks have separate containers to hold each type of recyclable. Other trucks put all the recyclables together and the trash company separates them at a processing facility.

Curbside pickup isn't the only way to collect recycling. Here are a few other methods for disposing of your recyclable goods:

- Drop-off centers are places where you can bring many different kinds of recyclable goods. These facilities accept everything from paper and bottles to hazardous materials, such as batteries or paint. Although drop-off centers save money on pickup

At the Great Minnesota eCycling Event, people drop off their old TVs, computers, VCRs, and other electronics equipment to be recycled instead of thrown in a landfill.

costs, people like you have to make the effort to bring your recyclables there.

- Deposit/refund programs pay people a cash refund (usually 5 or 10 cents) for dropping off their bottles or aluminum cans. People can either visit a bottle collection facility or use a reverse vending machine (often found at supermarkets), which accepts empty bottles and cans and dispenses cash (or a refund receipt) in exchange. Not every state requires that companies put a refundable deposit on bottles and cans. Only about eleven states currently have these "bottle bills."

- Buyback centers pay cash for your old aluminum, scrap metal, glass, plastics, and newspapers.

- Volunteer pickup is when community organizations (such as the Boy Scouts or a church group) raise money by collecting recyclable goods and bringing them to bottle collection facilities or buyback centers.

As these programs illustrate, there are many ways to handle recycling. Some of them even reward you with money for your efforts.

FROM CURB TO NEW PRODUCT: HOW ITEMS ARE RECYCLED

No matter where they are picked up, when recyclable goods arrive at a processing facility, they are separated. From there, each item takes a different journey, similar to one of the following:

Aluminum cans are the most recycled type of container. In 2006, the United States recycled 51.9 billion cans, according to the Aluminum Association. Aluminum recycling is very efficient because it uses 95 percent less energy than making new aluminum. To recycle aluminum, used cans are crushed and the labels are burned off in a very hot oven called a furnace. Then, the aluminum is melted, and the molten aluminum is poured into bar-like shapes called ingots. These ingots are rolled into flat sheets and stretched very thin. Most old aluminum cans are used to make new aluminum cans, but they also can be made into objects like car parts, tin foil, and pie plates.

Paper is sorted by color, weight, type (newspaper, boxes, writing paper), and whether it has already been recycled or not. Then, it goes into a big, hot bath filled with chemicals and water, which breaks the paper down into fibers. After a while, the paper turns into a soup-like mush called pulp. The pulp goes through a filter, which removes staples and glue. Chemicals pull the ink off the paper in a process called de-inking. The pulp may also be bleached with hydrogen peroxide or chlorine to turn it white. Then it is rolled through a big machine into flat sheets. The sheets are heated, dried, pressed, and wound onto a giant roll. Finally, the roll is cut into sheets of new paper, which is used to make newspapers, tissue, boxes, packaging, and more paper.

Plastics aren't recycled as often as paper and aluminum because they are very inexpensive to produce. Only about one out of every four bottles is recycled, according to the Washington State Department of Ecology. When plastic is recycled, it is first sorted by color and type of plastic. You can tell what type of plastic it is by looking at the bottom of the bottle or package. The recycling code number is inside a triangle made of arrows (the universal recycling symbol). After the plastic is sorted, it is chopped up into tiny pieces and melted. Recycled plastics can be made into things like new furniture, insulation, construction materials, lunch trays, and egg cartons.

RECYCLING CODES FOR DIFFERENT TYPES OF PLASTICS

The next time you get a plastic bottle, turn it over and look for the universal recycling symbol and the code on the bottom. Then, compare it to the chart below to see what type of plastic was used to make it.

PLASTIC RESIN CODES

1 PETE	2 HDPE	3 V	4 LDPE	5 PP	6 PS	7 OTHER
Polyethylene Terephthalate	High Density Polyethylene	Vinyl	Low Density Polyethylene	Polypropylene	Polystyrene	Other
soda bottles	milk, water and juice jugs	clear food packaging	bread bags	ketchup bottles	meat trays	ketchup
water bottles	detergent bottles	shampoo bottles	frozen food bags	yogurt and margarine tubs	egg cartons	3 & 5 gallon water bottles
shampoo bottles	yogurt and margarine tubs		squeezable bottles (mustard, honey)		cups and plates	some juice bottles
mouthwash bottles	grocery bags					
peanut butter jars						

These codes identify the type of plastic used to make different products. By looking at the code, you can find out whether or not an item can be recycled.

These codes don't necessarily mean that all the plastics can be recycled—they only identify the type of plastic. Types 1 and 2, for example, are widely used in container form. Type 4 is sometimes used in bag form. Types 6 and 7 have almost no recycling potential and are harmful to your health and the environment, according to a report released by the U.S. Government Accountability Office.

Glass can be recycled, but not all types of glass can be used again easily. Window, mirror, and lightbulb glass is too expensive to recycle, so usually only glass from jars and bottles is recycled. When glass bottles are collected, they are sorted by color and ground into very tiny bits called cullet. Other substances, such as limestone, sand,

and soda ash are added to the cullet, and then the mixture is melted down and formed into new glass. Most glass bottles contain at least 25 percent recycled glass, according to the Washington State Department of Ecology.

Steel is recycled from old cars and buildings. U.S. laws require that all steel used in manufacturing contains at least some recycled steel. Old cars and building materials are shredded, melted, and then rolled into huge sheets or coils. Everything from railroad ties to new cars can be made from recycled steel. Even scrap steel from the World Trade Center has been recycled. After terrorists struck the Twin Towers on September 11, 2001, steel from the two buildings was melted down

The bow of the USS *New York* was made from approximately 7.5 tons (6.8 metric tons) of steel taken from the World Trade Center after the terrorist attacks of September 11, 2001.

and forged to make the bow (forward part) of the military ship the USS *New York*.

Tires are difficult to recycle because the rubber used to make them goes through a process called vulcanization. This process makes tires bouncy and springy, but it also makes the rubber harder to melt down. To recycle tires, companies have to first shred them into strips. Powerful magnets pull out any metal that is inside the tires. Then, the strips are fed into a machine that grinds them into small pieces. Used tires can be used in manufacturing items like playground surfaces, doormats, doorstops, and garden hoses.

Batteries often contain toxic chemicals such as lead, mercury, and nickel, which can be very dangerous if they are released into the environment. That's why most states require people to recycle batteries—especially car batteries. Different types of batteries have their own recycling process. Lead batteries, for example, are placed into a machine called a hammermill, which crushes them into pieces about the size of a nickel. The pieces are placed into a tank, where they are separated by weight. The heavier pieces (lead) sink to the bottom, while the lighter pieces (rubber and plastic) float to the top. Each of these materials is treated separately. Lead is melted in furnaces and poured into molds. It is cooled and then shipped to manufacturers to be used in new batteries. Plastic, often used in car battery covers, is recycled just like other types of plastic. A special chemical is added to battery acid to turn it into a harmless mixture of water and salt. Once it has been tested to make sure it is safe, the recycled acid is sent into the public sewer system.

Electronic waste, or e-waste, is becoming a growing area of the recycling industry. Consider all the electronics you have in your home: cell phones, personal music players, DVD players, TV sets, video games, and more. Now, multiply those items by the millions of households in the United States. Finally, consider how often you replace each of your electronics and buy new products. A statistic from the organization

Many stores will accept used cell phones and other electronics, known as e-waste, for recycling at selected drop-off sites.

Keep America Beautiful reveals that the average person gets a new computer every three to five years. That means hundreds of millions of computers turn into junk every year. You can't just throw out electronics because many of them are made with toxic materials, such as lead, nickel, and mercury. A single computer monitor can contain up to 8 pounds (4 kg) of lead. You can donate electronics that still work to schools and nonprofit organizations. Manufacturers will often accept broken electronics. They will take the electronics apart and use many of the pieces (including batteries, plastic handsets, and circuit boards) to make new products.

RECYCLING LANDFILLS

Landfills are usually thought of as places to throw away trash, yet they actually can be an important part of the recycling process. When the organic waste in landfills decays, it releases methane, a type of natural gas. Companies can capture that gas and use it for energy.

Sometimes, an entire landfill can be recycled. When people visit the Mountain Gate Country Club near Los Angeles, California, all they see is an elegant golf course surrounded by expensive homes. Yet, underneath all the lush grass is a landfill. The entire country club was built on top of solid waste deposits. Not only has the land been recycled, but also the methane gas created by the decomposing waste is being captured and used to power buildings at the nearby University of California, Los Angeles.

HOW ITEMS ARE REUSED

Reusing old items is another way to prevent them from winding up in landfills. The idea behind reuse is to find a new home—or purpose—for something that's no longer needed. Some of the ways companies and individuals reuse include the following:

- Using building scraps and unused construction materials to make houses for people who can't afford a place to live. Some builders use recycled denim for insulation and reclaimed wood for flooring. Some architects are even converting old steel shipping containers into housing.

Goodwill Industries and similar charities accept donations of used clothing and household goods, which they sell inexpensively to people in need.

- Donating unused canned goods to food banks, where they go to feed the hungry.
- Giving used furniture to Goodwill or a thrift store so other people can buy it at a low cost.
- Providing unused art supplies to schools that lack supplies.
- Dropping off clothing at a consignment store so you can make money on items you no longer use.
- Donating gently used or never-used toys to churches or organizations that give these gifts to less fortunate families during the holidays.

Reuse is a win-win situation. It helps people and companies that no longer need their old products clear out clutter while providing much-needed goods to people who can really use them.

TEN GREAT QUESTIONS TO ASK A SCIENCE TEACHER

 1 What types of items can I recycle?

 2 How does recycling help the environment?

 3 Why are landfills a problem?

 4 What new products are made from recycled items?

 5 Is recycling really worth the time and effort?

 6 Which types of recyclable items contain toxic substances?

 7 How can I reuse things I have around the house?

 8 How can we reuse supplies or other objects in the classroom and at school?

 9 Can we plan a field trip to a local recycling center?

 10 What are some recycling projects I can do at home?

CHAPTER 3

Recycling and Reuse Today

Recycling and reuse may seem like new ideas, but they've been around for millions of years. In fact, the biggest recycler and reuser of all is nature. When plants and animals die, they are absorbed into the ground or oceans. Those decomposing bodies provide nourishment for the next generation of plants and animals.

Humans got into the act of recycling much later. Around 200 BCE, the Chinese found a new use for old fishing nets—they created the world's first recycled paper. The ancient Mayas used to recycle entire buildings. Instead of tearing down old buildings and starting from scratch, they simply added on to existing structures.

The U.S. government began to get actively involved in recycling during the twentieth century. During World War I (1914–1918) and World War II (1939–1945), soldiers needed a constant flow of weapons, clothing, and other goods at the European front. The American government encouraged people to collect everything from newspapers to aluminum to

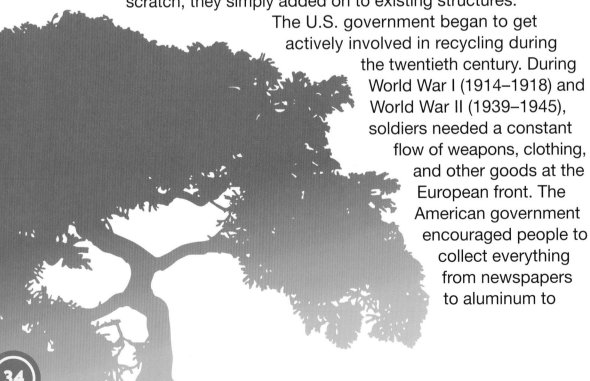

During World War II, the American government encouraged people to save whatever they could to help the war effort. This Philadelphia Salvage Committee poster urged people to save for reuse.

TRASH BARGE

The name *Mobro* might not sound familiar to you, but it was once the most famous trash barge in the world. On March 22, 1987, the *Mobro* left New York Harbor carrying 3,200 tons (2,903 metric tons) of garbage. It traveled 6,000 miles (9,656 km) along the East Coast of North America, trying to find a place to dump its load. The *Mobro* went as far south as the country of Belize in Central America, but no port would accept its garbage. Finally, after six months of sailing, the barge came back to Brooklyn, New York, where its trash was incinerated. Newspapers and television stations covered the *Mobro*'s journey. This huge media event called attention to the nation's growing trash problem, and it helped make people realize the importance of recycling.

help the military effort. The motto during World War I was, "Don't Waste—Save It."

In 1965, the U.S. population had grown to more than 193 million and the number of consumer goods being used—and thrown away—rose. That year, the government passed the Solid Waste Disposal Act to help manage the country's growing trash problem. In 1976, the act was renamed the Resource and Conservation and Recovery Act. It set up national standards for treating and storing wastes. The act also emphasized conservation, in part by encouraging people and companies to recycle.

The recycling revolution really got going on April 22, 1970, when Senator Gaylord Nelson of Wisconsin started the very first Earth Day

The first Earth Day celebration was held on April 22, 1970. This annual event has helped promote recycling and environmental awareness.

celebration. One day each year, Americans gather for festivals and rallies to promote environmental awareness and find ways to protect the planet. Recycling plays a big part in this annual celebration, which helped launch the U.S. Environmental Protection Agency (EPA). Today the EPA oversees recycling efforts across the country and sets goals for recycling more of the trash people use. In 1988, the EPA announced a five-year goal to recycle 25 percent of trash. By 2008, Americans were recycling just over 30 percent of their garbage.

The United States still does not have a national recycling law. However, many states have established their own laws requiring residents to recycle. Some states have passed laws that pay people back a 5- or 10-cent deposit when they return bottles and cans. Eleven states—including California, Oregon, Connecticut, Massachusetts, Michigan, and New York—currently have these bottle and can recycling laws. Some cities and states have laws that prevent people from throwing away recyclable items. If you live in New York City, for example, you can be fined $25 for mixing recyclables in with your regular trash.

LOCAL COMMUNITIES MAKING A DIFFERENCE

Although recycling isn't the law across the country, local communities, religious organizations, and schools have stepped in and started programs of their own. The following are just a few of the community-based recycling programs that have made a difference.

School lunch tray recycling program. Many schools recycle paper and aluminum cans, but one Georgia school district took recycling a step further. When the Gwinnett County Public School system discovered that it was sending more than sixteen million polystyrene plastic lunch trays to landfills each year, it came up with a plan to collect and recycle the trays. In its first year, the program recycled more than six million lunch trays.

RecycleBank. In 2004, Ron Gonen founded RecycleBank in Philadelphia, Pennsylvania. The idea was to encourage people to recycle

Almost anything can be reused—even prom dresses. These designer prom dresses are being donated to high school girls who need them.

by offering them something valuable in return. When people recycle through RecycleBank, they earn points, which can be used to buy products at hundreds of different stores.

Recycling lottery. In the city of Rockford, Illinois, residents have a good reason to recycle—a big cash prize. One homeowner's trash is selected by chance each week and checked. If no recyclable items are found, the homeowner wins $1,000. If the person selected doesn't win, the lottery goes up to $2,000 the next week, $3,000 the following week, and so on. After a few months of launching this lottery, Rockford had a 400 percent increase in recycling!

Putting computers to good use. One company is doing something good with old computers. The Assistive Technology Exchange Network in Illinois collects computers from companies that no longer need them and donates them to children who have disabilities.

Giving prom dresses a second life. When Shauna Cole, director of the Teen Program at the Reston Community Center in Reston, Virginia, found out that many local high school girls couldn't afford prom dresses, she came up with the perfect solution. She started collecting prom dresses, shoes, jewelry, and makeup from people all over northern Virginia. The center's Prom Dress Giveaway collected five hundred dresses in its first year, giving hundreds of girls an unforgettable prom night.

RECYCLING AROUND THE WORLD

The United States is getting better about recycling, but other countries have come up with their own unique ways of making sure their trash doesn't wind up in landfills.

- In Switzerland, recycling is free, but trash is expensive. Supermarkets collect bottles, towns pick up paper and yard waste, and aluminum and tin can be dropped off at local depots at no charge. Yet, for each bag of trash

In Sweden, residents and businesses in every neighborhood are required to sort their recyclables into different color-coded containers.

residents put out for collection, they must pay 1.50 in Swiss francs (about $1.30).

- In Bali, Indonesia, people don't just take used bottles to a recycling center. They wash them by hand and then refill them with drinks themselves.
- In East Africa, people use old cans and other used items to make toys and fix the roofs on their homes.
- In Germany, people collect their recyclables in at least five different bins: packaging, paper, glass, food waste, and other types of trash. Each bin has a different color to make sorting easy.

SCIENCE MEETS RECYCLING AND REUSE

The idea behind recycling is pretty basic: put your used bottles, cans, and paper out by the curb and a truck picks them up to be taken to a processing plant. Traditional recycling is just one part of the picture, though. As technology gets more advanced, science is coming up with new ways to recycle and reuse.

One of those ways is to produce products that will be biodegradable (able to break down) more quickly in landfills. It's an idea called green chemistry. An example is that of plastic. Most plastics are not very biodegradable. If you throw out a plastic cup, it can sit in a landfill for hundreds of years without breaking down. One company, SpudWare, is helping to solve that problem by making plastic out of something people normally eat: potatoes. The company's forks and knives are made from 80 percent potato starch and 20 percent soy oil, which break down in about six months. Utensils can also be made from corn, cassava, and sugar cane.

Today many companies start thinking about how their products might break down before they even build those products. The idea is called design for disassembly. For example, before home builders construct a new house, they look at all the pieces that will go into that house—windows, doors, flooring, carpets, and so on. The builders try to use the fewest number of parts and materials possible, and then they figure out how each of the pieces used to make the house can one day be reused, recycled, or safely thrown away.

Another new idea in recycling is to turn garbage into something usable—such as fuel or fertilizer. A few companies are testing a method for turning wood chips into ethanol, a type of alcohol that can be used to power cars. Almost any type of material containing hydrogen, carbon, and oxygen—from garbage to leftover farm crops—could potentially be turned into fuel because they release energy when burned. A few seafood companies are even turning old fish parts into fertilizer. With creative ideas like these, the recycling possibilities are virtually endless.

A factory worker in a Minnesota mill supervises the weaving of fiber into a blanket. The fiber pictured here was made from corn-based plastic.

MYTHS AND FACTS MYTHS AND FACTS
MYTHS AND
FACTS
MYTHS AND
FACTS
MYTHS AN
FACT.
MYTHS
AND FACTS
MYTHS AND
FACTS

MYTH: Recycled items are always worth less than the original products.

FACT: Although you may get only 5 cents for the bottle of soda that cost you $1, some recycled goods are actually worth more than what was paid for them. For example, some artists charge thousands of dollars for paintings and sculptures that they made out of garbage.

MYTH: Dumping trash in landfills costs less than recycling.

FACT: Sometimes, that is true because it can cost more to pick up, separate, and process recyclables than it costs to dump them in a landfill. However, when recycling programs are done right, they can actually be cheaper than landfills. It costs a lot less to recycle most old items than it costs to find the raw materials and make a whole new product, according to the EPA. The more people recycle, the less expensive recycling programs will become.

MYTH: You can't reuse disposable items.

FACT: Anything can be reused, even products that were designed to be used only once. For example, once you've finished your soda, you can save the can, cut it up, and make a piece of art. Or, you can string together the soda tabs to make a belt. Instead of throwing away your empty Styrofoam cup, you can plant flower seeds in it.

CHAPTER ④

How You Can Recycle and Reuse

If you're like most kids, your house is full of stuff. You have books you no longer read, CDs from bands you don't like to listen to anymore, clothes you've outgrown, and toys you no longer play with. What should you do with all that stuff?

Just about everything in your home can be recycled, with a little creativity. The following are a few resourceful ways to recycle and reuse the things you no longer need.

E-waste. Most electronics, including TVs, computers, stereos, VCRs, and cell phones, can be reused or recycled. It's important to be careful with any type of e-waste, though, because these products can often contain toxic chemicals like lead, mercury, and arsenic. These chemicals can be dangerous both to the environment and to the people and animals that live in it.

If you have an electronic device that still works, consider donating it before you throw it away.

American children donated notebooks and other books to these children in Afghanistan to help supply classrooms and libraries.

Give it to a local school, religious organization, or charitable agency like Goodwill. Most computer manufacturers, including Dell and Hewlett-Packard, will take back old computers once you're done using them. Stores, including Office Depot, Circuit City, Best Buy, and Staples, will accept used toner and ink cartridges. A few nonprofit companies, such as the Wireless Foundation, will accept old cell phones, batteries, and chargers. They will fix the phones and give them to people who need them.

Books, newspapers, magazines, and junk mail. Donate your used books to a local library or school. Most libraries will give you a receipt that your parents can use to save money when they file their taxes. Or, you can give your books to an organization like the International Book Project, which donates them to libraries and schools in other parts of the world. You can even make money off your old books by selling them to a used bookstore or putting them up for sale (with an adult's help) online, for example, on Amazon.com or eBay.

Your trash company should pick up old newspapers, magazines, and mail in your recycling bins. To reduce the amount of mail that needs to be thrown away, ask your parents to call and cancel delivery of any magazines and catalogs they don't read, and have them ask junk mail companies to take your family's name off all mailing lists.

There are many ways to reuse old newspapers. You can use them to wrap fragile items before shipping them, clean your windows with them (they work like paper towels), or line the table with them while you are doing art projects.

Batteries. You have to be careful when disposing of batteries because they contain dangerous acid and other chemicals. Never throw any batteries in the trash. Bring rechargeable batteries to stores like Office Depot and RadioShack, which will take them back. You can take alkaline batteries to your local household hazardous waste facility, or to a post office that offers battery disposal services. Many auto parts stores and service centers will accept car batteries for recycling.

Automobiles. Nothing is too big to recycle, including your mother or father's mammoth sports utility vehicle (SUV). Companies like the American Diabetes Association and Habitat for Humanity will take cars as gifts, as well as trucks, motorcycles, boats, Jet Skis, and recreational vehicles (RVs), even if the vehicles no longer run. Plus, your parents will get a big tax deduction for their donation.

Food scraps. Instead of throwing your banana peels down the disposal, why not put them to good use? Starting a compost bin in your backyard will not only help get rid of food scraps, but it will also generate enough fertilizer to keep your lawn lush and green. In a compost bin, bacteria, fungi, or worms eat the waste, break it down, and release an organic fertilizer called compost.

To make a compost bin at home, put shredded newspaper, peat moss, or sawdust into a bin. Add water to moisten it slightly. Then, put about 2 pounds (0.9 kg) of red worms or red wigglers into the bin. Place your garbage on top. You can include refuse like coffee grounds, fruit rinds, vegetable scraps, tea bags, eggshells, and lawn clippings. Cover the bin lightly, but leave a little bit of air so the worms can breathe. After a few weeks, you should be able to "harvest" the compost to use as fertilizer.

Other things you can recycle. Don't limit yourself to the items mentioned here. You can find a new use for just about everything in your home. Take your used shoes and clothes to the Salvation Army or one of the many other charity organizations that collect household items. Donate your broken crayon pieces to Crazy Crayons, an organization that makes new crayons out of old ones. Give your old eyeglasses to the Lions Clubs or bring them back to the store where you bought them; most stores that sell glasses will donate them for you to those in need. You can even recycle wire hangers by bringing them back to your dry cleaner.

CREATIVE WAYS TO REUSE DISPOSABLE ITEMS

Is your house getting too cluttered? Try these simple projects for using objects you have around your house:

1. Take the cover out of your CD case and remove the CD so you are left with just the clear plastic case (you also can reuse the CDs as drink coasters). Insert one of your favorite photographs, and turn the empty CD case into a desktop picture frame.

These high school students in Shreveport, Louisiana, collected recyclable plastic bottles that will be made into a type of fleece that can be woven into jackets for children in need.

2. Rinse out yogurt containers. Add soil and plant seedlings to create your own windowsill herb garden.
3. Tape patterned gift wrap around the outside of a potato chip can and use the can to hold pens and pencils on your desk.
4. Cut out pictures from last year's calendar to make bookmarks. To give your bookmarks more weight, cut strips from used greeting cards and tape them to the back.
5. Put used dryer sheets in your dresser drawers to keep your clothes smelling fresh.
6. Tape a magnet to the back of an empty raisin box and use it as a refrigerator magnet. You can also use empty candy boxes, juice boxes (just rinse them out first), bottle caps, and mint tins.
7. Paint bottle caps red and black and make your own checkers game. Draw red and black squares on a piece of cardboard to use as the checkerboard.
8. Save used jars or coffee containers. Fill them with homemade cookies or candy, decorate them with a bow, and give them as gifts to friends and relatives.
9. Cut up used pieces of paper into small squares. Turn them over so the clean side is showing, and put them in a pile near the phone for writing messages.
10. Use paper and plastic grocery bags instead of trash bags to line your wastebaskets.
11. Punch two holes about 2 inches (5 cm) from the bottom of an empty plastic soda bottle and insert a pencil to serve as a perch for birds. In your homemade bird feeder, add a few small holes slightly above the pencil so the birds can get to the seed. Fill the soda bottle with birdseed. Use twine to tie the bird feeder to a tree branch.
12. Wash an old shower curtain with hot water and disinfectant, remove the rings, and use it as a tablecloth for picnics.

STRANGE STRUCTURES MADE FROM RECYCLING

There are a lot of easy ways to recycle things you have around the house, but some people have taken the idea of recycling much further. They have come up with some pretty unique inventions made from recycled trash.

Plastic bridge. In 2002, researchers from Rutgers University in New Jersey built a bridge 42 feet (12.8 m) long out of a type of plastic found in disposable cups and milk jugs. The bridge is a fire access route that spans the Mullica River, and it's strong enough to hold a 36,000-pound (16,329 kg) fire truck.

Recording road. What do you do with one million unsold copies of a music CD? When a British record company found itself stuck with a warehouse full of CDs by a pop singer, it crushed them up and shipped them to China, where they were used in resurfacing roads.

Trash house. In Rio de Janeiro, Brazil, construction worker Luiz Bispo found a unique use for his trash—he built his home out of it. Bispo said the entire house cost him only $170. He made most of his house from construction waste and furniture that he had found at dumps. Bispo's house floats on hundreds of empty plastic bottles in a channel.

13. Stuff a rubber glove with ice to make an ice pack. Keep it in your freezer to use when you get injured.
14. Wrap your gifts using the Sunday comics instead of decorative wrapping paper purchased at a store.

These students in Delaware learn how to compost their trash using worms. Composting yard waste is one way to cut down on the use of landfills.

ENCOURAGING YOUR SCHOOL TO RECYCLE AND REUSE

Don't wait for your teachers or principal to set up a recycling program—take the first step yourself. You'd be surprised at how much you can accomplish if you set clear goals and stick to them.

If your school does not recycle, write up a plan—a list of activities your school can start doing now. The ideas that follow are some that you might include on your list.

- Schedule a meeting in which all the teachers can get involved and share their ideas for a school-wide recycling program.

- Establish a team made up of students and teachers to lead the recycling effort. One group of students can serve as "recycling guides," visiting classrooms and making sure that the program is running smoothly. Reward students who participate with a special party at the end of the year.
- Decide what items you will recycle or reuse. Your list might include paper, bottles and cans, books, school and art supplies, and sports equipment.
- Put recycling bins (especially for paper and aluminum) in every classroom. Clearly label the bins and put them right next to the trash cans so it's easy for your classmates to recycle.
- Get your fellow students excited about recycling by setting goals. If you have deposits on bottles and cans in your state, one idea is to have students aim to collect ten thousand bottles and aluminum cans to raise money for a fun school project. Or, you could encourage everyone to bring in a bag of old clothes and host a school yard sale.
- Take your list straight to the top of your school. Set up a time to present your ideas to the principal and other school administrators. Ask them how you can get your teachers and classmates involved and set a date to launch your program.

Once you have everyone in your school involved in recycling and reusing, start working on your parents and siblings. Make sure you have recycling bins at home and that everyone in your house is using them. Sit down for a family meeting to come up with new ideas for recycling and reusing. Encourage your parents to spread the word to their friends and coworkers. The more people you get involved in recycling, the bigger the positive influence you can have on your world.

GLOSSARY

bacteria Tiny living organisms that cannot be seen with the eye alone. Some bacteria cause illness or rotting, but others are helpful.

biodegradable Describes a substance that can be broken down by air, water, and bacteria.

compost A process by which organic materials like food scraps and yard waste are broken down. Compost can be used as fertilizer.

conservation The wise use and protection of natural resources.

cullet Crushed, used glass that is added to make new glass.

decompose To decay or break down into parts.

design for disassembly Making products using materials that can be easily recycled, and designing them so they can be taken apart easily and their impact on the environment can be reduced.

ethanol A type of alcohol that can be used to power automobiles and other vehicles instead of, or in addition to, gasoline.

e-waste Televisions, computers, video game players, and other discarded electronics equipment.

global warming An increase in the earth's temperature caused by the release of greenhouse gases (such as carbon dioxide) into the atmosphere.

hazardous Harmful or dangerous.

incineration The process by which garbage is burned.

incinerator A furnace or container used to burn waste at very high temperatures.

landfill An area of land where trash and garbage are buried between layers of earth and special barriers, to help slow leachate from running into the groundwater.

leachate Liquid that collects and runs off from the garbage in landfills.

methane A colorless, odorless, and very flammable gas that is produced as garbage decays in landfills.

pulp A soggy mass of used paper, water, and chemicals that is used to make new paper.

recycling Processing used items to turn them into something new.

reducing Using less of something.

reusing Using a product more than once.

toxic Poisonous.

vulcanization The chemical process that involves heating rubber and sulfur together to harden rubber, which also makes it difficult to recycle.

FOR MORE INFORMATION

GrassRoots Recycling Network
P.O. Box 282
Cotati, CA 94931
(707) 321-7883
Web site: http://www.grrn.org
This network of activists and recycling professionals is dedicated to the
idea of zero waste—not wasting any resources.

National Recycling Coalition
805 Fifteenth Street NW, Suite 425
Washington, DC 20005
(202) 789-1430
Web site: http://www.nrc-recycle.org
The members of this nonprofit group are devoted to finding new ways
to recycle and reuse waste.

National Resources Defense Council
40 West 20th Street
New York, NY 10011
(212) 727-2700
Web site: http://www.nrdc.org
This grassroots environmental organization helps promote green living
and finds other ways to protect the environment.

Recycling Council of Ontario
215 Spadina Avenue, #407
Toronto, ON M5T 2C7
Canada
(416) 657-2797

Web site: http://www.rco.on.ca
This nonprofit agency in Canada teaches people how to reduce waste
and how to get rid of waste in a more environmentally friendly way.

Recycling and Environmental Action Planning Society (REAPS)
P.O. Box 444
Prince George, BC V2L 4S6
Canada
(250) 561-7327
Web site: http://www.reaps.org
REAPS helps people learn about recycling, composting, and many
other environmental issues.

U.S. Environmental Protection Agency (EPA)
Ariel Rios Building
1200 Pennsylvania Avenue NW
Washington, DC 20460
(202) 272-0167
Web site: http://www.epa.gov
This agency of the U.S. government works to protect human health
and the environment.

WEB SITES

Due to the changing nature of Internet links, Rosen Publishing has
developed an online list of Web sites related to the subject of this book.
This site is updated regularly. Please use this link to access the list:

http://www.rosenlinks.com/gre/recy

FOR FURTHER READING

Barraclough, Sue. *Recycling Materials* (Making a Difference). Mankato, MN: Sea to Sea Publications, 2007.

Cothran, Helen, ed. *Garbage and Recycling: Opposing Viewpoints*. Farmington Hills, MI: Greenhaven Press, 2003.

Ganeri, Anita. *Something Old, Something New: Recycling* (You Can Save the Planet). Portsmouth, NH: Heinemann, 2005.

Hewitt, Sally. *Waste and Recycling* (Green Team). New York, NY: Crabtree Publishing Company, 2008.

Inskipp, Carol. *Reducing and Recycling Waste* (Improving Our Environment). Strongsville, OH: Gareth Stevens Publishing, 2005.

Levete, Sarah. *Rot and Decay: A Story of Death, Scavengers, and Recycling* (Let's Explore Science). Vero Beach, FL: Rourke Publishing, 2007.

Orme, Helen. *Garbage and Recycling* (Earth in Danger). Kent, England: BearportPublishing, 2008.

Silverman, Buffy. *Recycling: Reducing Waste*. Portsmouth, NH: Heinemann, 2008.

Spilsbury, Louise. *A Sustainable Future: Saving and Recycling Resources*. Chicago, IL: Raintree, 2006.

Thomson, Ruth. *Rubber* (Recycling and Re-using Materials). North Mankato, MN: Smart Apple Media, 2006.

The Aluminum Association. "U.S. Aluminum Can Recycling Steady in 2006." June 27, 2007. Retrieved August 1, 2008 (http://www.aluminum.org/AM/Template.cfm?Section=Home&TEMPLATE=/CM/ContentDisplay.cfm&CONTENTID=25033).

American Chemistry Council. "Plastic Packaging Resins." Retrieved September 9, 2008 (http://www.americanchemistry.com/s_plastics/bin.asp?CID=1102&DID=4645&DOC=FILE.PDF).

Benjamin, Daniel K. "Recycling Rubbish: Eight Great Myths About Waste Disposal." *PERC Reports*, Volume 21, September 2003. Retrieved August 4, 2008 (http://www.perc.org/articles/article224.php).

Blatt, Harvey. *America's Environmental Report Card: Are We Making the Grade?* Cambridge, MA: MIT Press, 2005.

California Environmental Protection Agency. "The Illustrated History of Recycling." Retrieved August 4, 2008 (http://www.p2pays.org/ref/26/25070.pdf).

Cothran, Helen, ed. *Garbage and Recycling: Opposing Viewpoints*. Farmington Hills, MI: Greenhaven Press, 2003.

Deneen, Sally. "How to Recycle Practically Anything." *E—The Environmental Magazine*, May/June 2006, Vol. 17, p. 26–32.

Ellick, Adam B. "Houston Resists Recycling, and Independent Streak Is Cited." *New York Times*, July 29, 2008. Retrieved July 10, 2008 (http://www.nytimes.com/2008/07/29/us/29recycle.html?n=Top/News/U.S./U.S.%20States,%20Territories%20and%20Possessions/Texas).

Energy Information Administration. "Energy Kid's Page." Retrieved July 10, 2008 (http://eia.doe.gov/kids/energyfacts/saving/recycling/solidwaste/recycling.html).

HowStuffWorks. "How Recycling Works." Retrieved August 1, 2008 (http://www.howstuffworks.com/recycling.htm).

Keep America Beautiful. "Waste Reduction." Retrieved August 1, 2008 (http://www.kab.org/site/PageServer?pagename= Focus_Waste_reduction).

Nakazawa, Liz. "A New Corn-Based Plastic Disappears Into the Dirt." September 4, 2003. Retrieved September 18, 2008 (http://www. csmonitor.com/2003/0904/p12s02-sten.html).

National Recycling Coalition. "Top 10 Reasons to Recycle." Retrieved July 10, 2008 (http://www.nrc-recycle.org/ top10reasonstorecycle.aspx).

Small, Meredith F. "Real-World Recycling Puts U.S. to Shame." Yahoo! News, August 23, 2008. Retrieved September 16, 2008 (http://news.yahoo.com/s/livescience/20080823/sc_livescience/ realworldrecyclingputsustoshame).

Tierney, John. "Recycling Is Garbage." *New York Times*, June 30, 1996. Retrieved August 4, 2008 (http://query.nytimes.com/gst/ fullpage.html?res=990CE1DF1339F933A05755C0A960958260).

U.S. Environmental Protection Agency. "Batteries." Retrieved August 5, 2008 (http://www.epa.gov/garbage/battery.htm).

U.S. Environmental Protection Agency. "Combustion." Retrieved September 4, 2008 (http://www.epa.gov/epaoswer/education/ quest/pdfs/unit2/chap4/u2-4_combustion.pdf).

U.S. Environmental Protection Agency. "eCycling." Retrieved August 5, 2008 (http://www.epa.gov/epaoswer/hazwaste/recycle/ecycling/ basic.htm).

U.S. Environmental Protection Agency. "Municipal Solid Waste Generation, Recycling, and Disposal in the United States: Facts

and Figures for 2006." Retrieved August 5, 2008 (http://
www.epa.gov/msw/msw99.htm).

U.S. Environmental Protection Agency. "Resource Conservation and
Recovery Act." Retrieved August 21, 2008 (http://epa.gov/
superfund/students/clas_act/haz-ed/ff_06.htm).

U.S. Environmental Protection Agency. "Reuse + Recycling=
Waste Reduction: A Guide for Schools and Groups." Retrieved
September 20, 2008 (http://www.epa.gov/epaoswer/osw/
students/school.pdf).

U.S. Government Accountability Office. "Electronic Waste: Harmful
U.S. Exports Flow Virtually Unrestricted Because of Minimal EPA
Enforcement and Narrow Regulation." Retrieved September 17,
2008 (http://www.gao.gov/htext/d081166t.html).

Wald, Matthew L. "Gassing Up with Garbage." *New York Times*, July 24,
2008. Retrieved September 18, 2008 (http://www.nytimes.com/2008/
07/24/business/24fuel.html?ref=business).

Washington State Department of Ecology. "Kids Recycle Page."
Retrieved July 10, 2008 (http://www.ecy.wa.gov/programs/
swfa/kidspage).

Zeller, Tom. "Recycling the Big Picture." *National Geographic*, January
2008, Vol. 213, pp. 82–87.

INDEX

ABOUT THE AUTHOR

Stephanie Watson is a regular contributor to several online and print health and science publications. She has written or contributed to more than two dozen books, including *Biotechnology: Changing Life Through Science* and *Critical Perspectives on Pollution* (Scientific American Critical Anthologies on Environment and Climate). She lives in Atlanta, Georgia.

PHOTO CREDITS

Cover, p. 1 © www.istockphoto.com/Andrejs Zemdega; p. 6 © Steve LaBadessa/Zuma Press; p. 10 © James Leynse/Corbis; pp. 12–13 courtesy of The Solid Waste Authority of Central Ohio (SWACO); pp. 14, 16, 27, 39, 43, 49, 52 © AP Images; p. 19 Tim Sloan/AFP/ Getty Images; p. 29 Tim Boyle/Getty Images; p. 31 © Noah Addis/ Star Ledger/Corbis; p. 35 Library of Congress Prints and Photographs Division; p. 37 Hulton Archive/Getty Images; p. 41 © Rob Schoenbaum/ Zuma Press; p. 46 Farzana Wahidy/AFP/Getty Images.

Designer: Nicole Russo; Editor: Kathy Kuhtz Campbell;
Photo Researcher: Amy Feinberg